I0484919

Police Radar Basics

Everything Every Driver, *and the Police,* should know about Traffic Speed Radar

Donald Sawicki

Police Radar Basics
© Copyright 2015 Donald S. Sawicki
version: 05012015

Contents

Chapter 1 - Radar Equipment   **1**

    1.1 - Radar Types   1

    1.2 - Radar Displays   4

    1.3 - Antenna Alignment   5

    1.4 - Radar Options   8

    1.5 - Power Source   10

    1.6 - Paperwork Traceability   11

Chapter 2 - Tuning Fork Test   **15**

    2.1 - Tuning Forks   16

    2.2 - Stationary Mode Test   17

    2.3 - Moving Mode Test   19

Chapter 3 - How It Works   **22**

    3.1 - Detection Range   22

    3.2 - Beamwidth   23

    3.3 - The Cosine Effect   24

        Radar Blind Zone   25

Chapter 4 - Operation   **26**

    4.1 - Daily Test and Use   26

    4.2 - Track History   27

    4.3 - Stationary Mode   28

    4.4 - Moving Mode Opposite Direction Traffic   30

    4.5 - Moving Mode Same-Direction Traffic   32

    4.6 - Fastest Target Mode   33

Chapter 5 - Limitations   **34**

    5.1 - Radar Accuracy   35

        Target and Patrol Acceleration Limits   35

    5.2 - Target Vehicle Selectivity   36

    5.3 - Multi-path Errors   37

    5.4 - Electromagnetic Interference   38

    5.5 - Transmit Turn-On Time   40

    5.6 - Patrol Speed Shadowing   41

    5.7 - Other Errors   43

Chapter 6 -- Theory   **44**

    6.1 - Radar Bands   44

    6.2 - Transmissions and Reflections   46

Radar versus Lidar   **49**

US Interstates   **50**

Police 10 Codes   **51**

Index   **58**

NOTES

NOTES

Chapter 1 - Radar Equipment

1.1 - Radar Types

Police radars come in two basic types, a **hand-held radar gun** or a **multi-unit** system **fix mounted** to the patrol car. Some hand-held radars have a mounting fixture in the patrol vehicle for fixed or hand held use.

HAND-HELD RADAR GUN

These radars are single piece units shaped somewhat like an oversized pistol and aimed in a similar manner. Most can only be used from a stationary position, some can be used from a moving patrol car using a mount fixture. Radars in a moving mode MUST have the antenna **mounted** in a fixed **straight** and **level** position. All hand-held moving mode radars come with a remote control.

Radars are powered by either the patrol car battery using a cord connected to the DC or cigarette plug, or internal batteries for cordless use. Some models have the option to run off of either internal batteries or the patrol car battery. Most radars with internal batteries can be recharged from the car battery (+12 Volts DC), and/or standard house current (110 Volts, 60 Hertz AC).

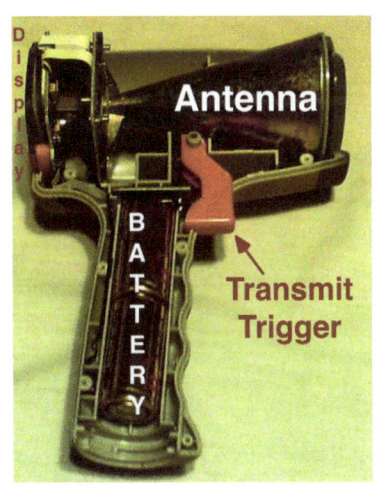

Hand-Held Radar Gun Internals

MULTI-UNIT FIX MOUNTED RADAR

Multi-unit radars mount to the patrol vehicle and operate from a stationary or moving mode.   These radars all have a front

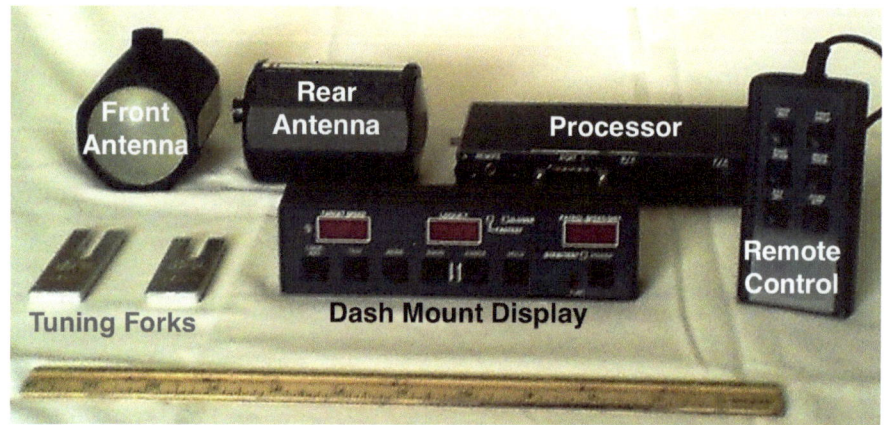

antenna, many also have an optional rear antenna.

Multi-Unit Radar Mounts to Patrol Vehicle

Rear antenna *side view*, front antenna *front view*.

Antennas can be mounted outside or inside the patrol car. Common inside mounting surfaces include the dashboard or windshield.   The antenna module includes the transceiver (transmitter and receiver) built-in.

Some models combine the computer and display into a single unit mounted to the dashboard.   Other units have a separate computer and display. The computer mounts under the seat or in the trunk.   The display module has a smaller dashboard footprint.

STATIONARY AND MOVING MODE **TUNING FORKS**

Tuning fork test should be performed daily, at the beginning and end of a shift, at a minimum.

Stationary radars come with at least one tuning fork. A vibrating tuning fork will induce a speed reading proportional to the tuning fork resonance frequency tone. The higher the tone or pitch, the higher the displayed speed. Each radar is assigned a tuning fork(s), tuned for that radar.

Typical Tuning Forks

Moving mode radars come with **two tuning forks**, with different tones. One of the tuning forks will induce a patrol car speed, the other fork is for the target speed. The patrol and target speed readings vary with radar *mode*; **opposite direction** (on-coming) traffic, or **same direction** (same-lane) traffic. Each radar is assigned a **set** of tuning forks, tuned for that radar.

1.2 - Radar Displays

All radars have a dedicated window for **target vehicle speed**. This speed is always the **strongest reflection**, usually the closest vehicle, but not always.

Many radars have a "**Locked**" speed window. The operator can lock (save) a reading while continuing to measure speed. This window is also used for (shared with) the **Fastest Speed** mode option. In this mode the radar can track 2 vehicles, the strongest reflection and the fastest vehicle.

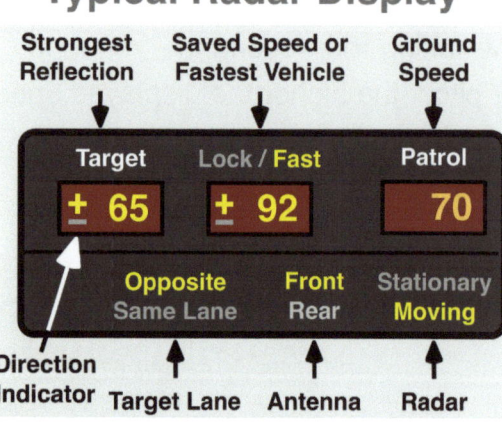

Some radars have a target vehicle *Direction Indicator*. A (**+**) indicates a positive Doppler and the target is getting closer, a (**-**) indicates a negative Doppler and the target is getting further away. Some models use arrows instead of (+) or (-).

All **Moving Mode radars** have a dedicated window for **patrol speed**.

RADAR SPEAKER

All radars can emit an audio tone proportional to the target speed, the higher the speed the higher the tone pitch. A strong steady stable reflection has a clean clear tone. A weak or unstable reflection, or interference, will have a noisy or unclear tone.

1.3 - Antenna Alignment

Antenna alignment is critical in moving mode. Front and rear antennas must be mounted **level** and pointed **straight** for **accurate** measurements.

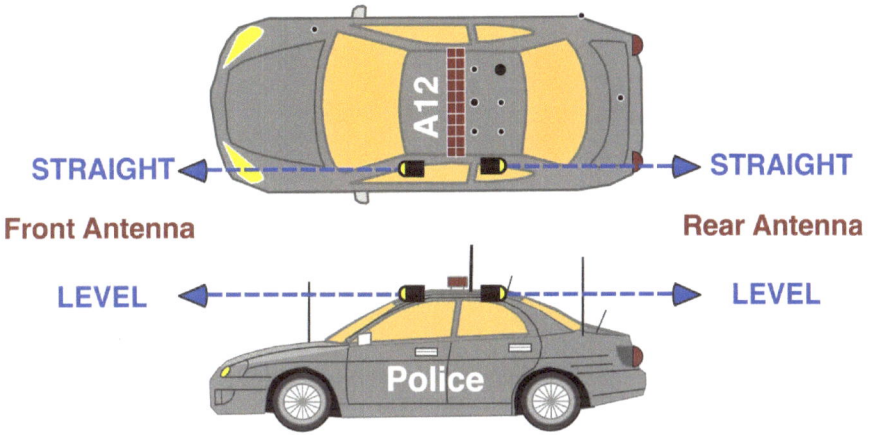

Front and **Rear Antennas** **MUST be** **Straight** and **Level**

Antenna beam must be **clear of obstructions**, higher mounting better.

Proper Alignment

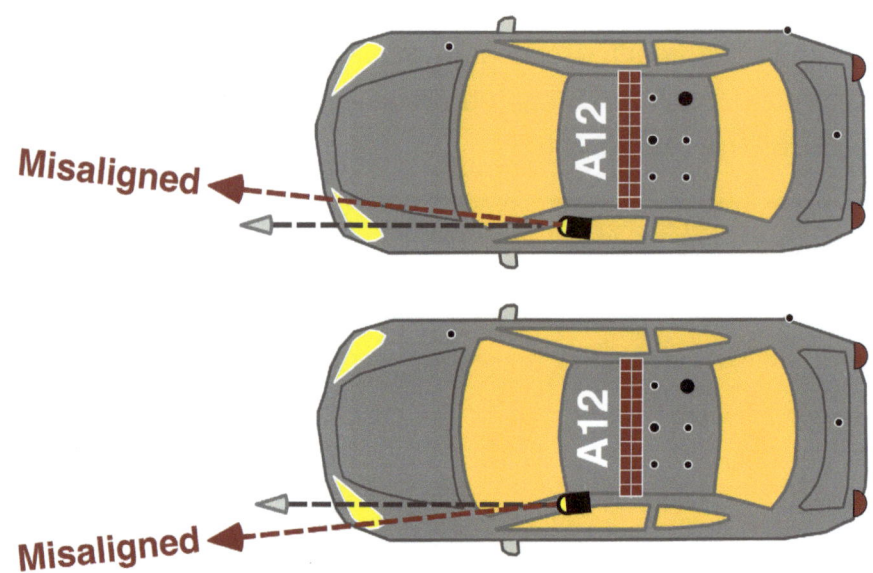

Misaligned Front Antenna

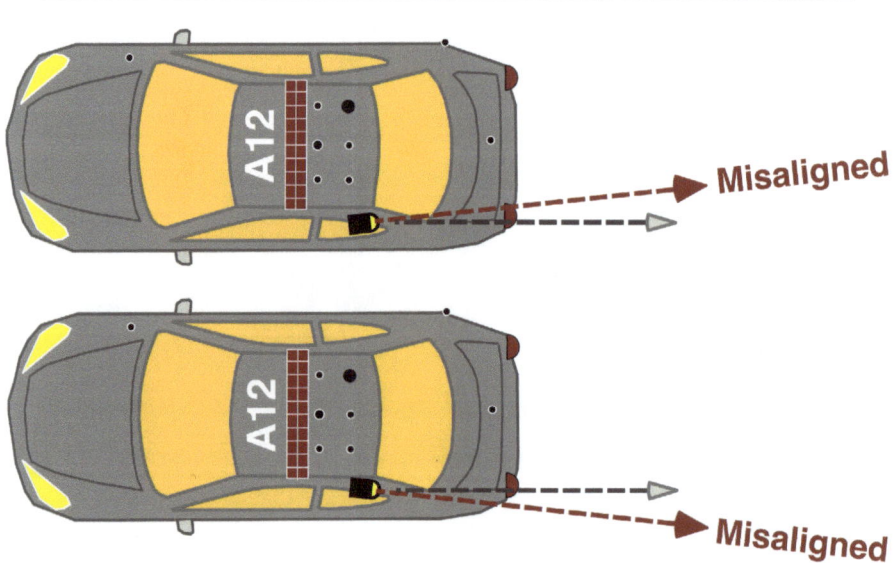

Misaligned Rear Antenna

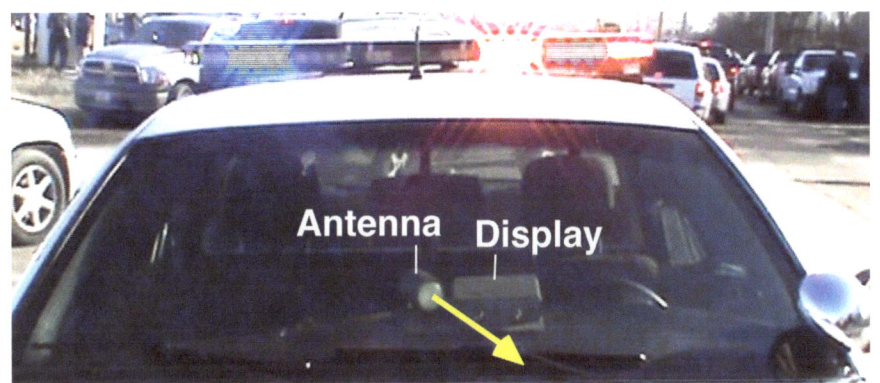

Grossly Misaligned Front Antenna

An antenna angled up (high) will miss targets, angled down (low) distorts the ground reflection. Good ground reflections are required in moving mode to calculate target speeds.

The greater the misalignment to the *right* or *left*, the greater the speed error (Cosine Effect). Moving mode radar is intended to measure ground speed directly in front of the patrol car, not off angle. Off angle ground reflections measure patrol speed low. A low patrol speed **translates directly** (1 to 1) to a **HIGH** target speed for most **radar moving modes** (with one exception, see below*). The **speed error** depends on the **angle error** and the **patrol speed**, and can vary from **a few** to **tens** of miles per hour (MPH). Larger angles and greater speeds result in larger speed errors. Also see Chapter 5.6 - *Patrol Speed Shadowing*, pages 41 - 42.

A patrol speed that is low by 10 MPH, results in a target speed high by 10 MPH.

* One exception is target speed measures low for **same-direction** *traffic traveling* **slower than the patrol** *car.*

1.4 - Radar Options

All radars have a front antenna, many multi-unit radars also have an additional optional **rear antenna**.

Connecting the radar to the **patrol car speedometer** (VSS -- Velocity Speed Sensor) greatly increases measured speed reliability, and eases operator workload in moving mode. Without the connection the operator **must** monitor the speedometer and compare it to the radar measured patrol speed, if the speeds do not closely match all speed readings are bad and must be rejected. This happens when the radar uses a ground reflection not directly in front of the patrol vehicle. It happens enough the phenomenon has a name -- "*Patrol Speed Shadowing*". A speedometer connection allows the radar to detect the error and automatically re-set ground track until a reliable ground speed is established.

A **dash camera** can display radar data (optional connection) that includes **Target** and **Patrol** Speed, **Antenna** (front or rear), and radar **Mode** (stationary, moving,*opposite* or *same-lane*).

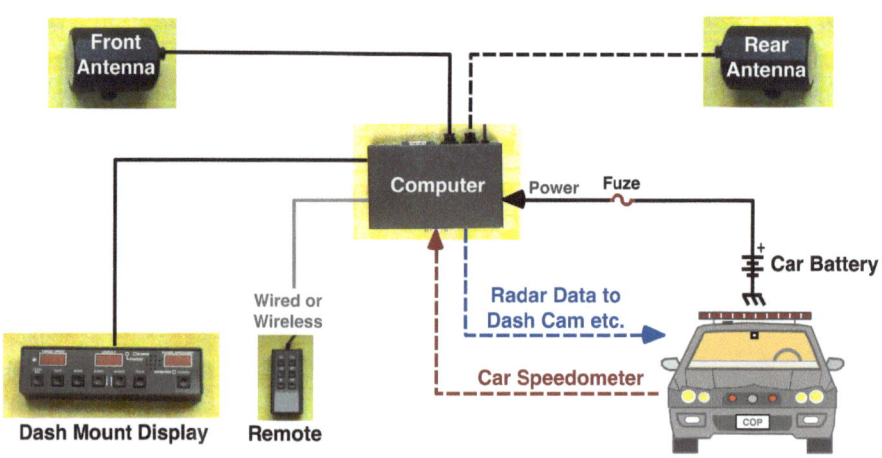

Radar Options (dashed lines)

8

Radar Fan Filter

The patrol car air circulation fan can be picked-up by a radar in some installations (not uncommon). The rotating fan will show up as a constant speed. The higher the fan speed, the higher the radar speed. Speeds vary with installation.

Vehicle
Fan on Low

Fan on High

Many radars have a **FAN FILTER** the operator can turn On or Off. The filter must be *turned-off* for the *tuning fork test*, then turned-on for use.

If the radar fan filter does not fix the problem, other techniques must be employed such as moving the antenna and/or shielding.

1.5 - Power Source

A patrol car's electrical system can cause interference through the power connection and produce radar false readings, and/or degrade detection range. Range degradation can be from slight to severe.

Noisy electrical systems may require additional filtering and/or re-wiring, or a separate (from patrol vehicle system) power source.

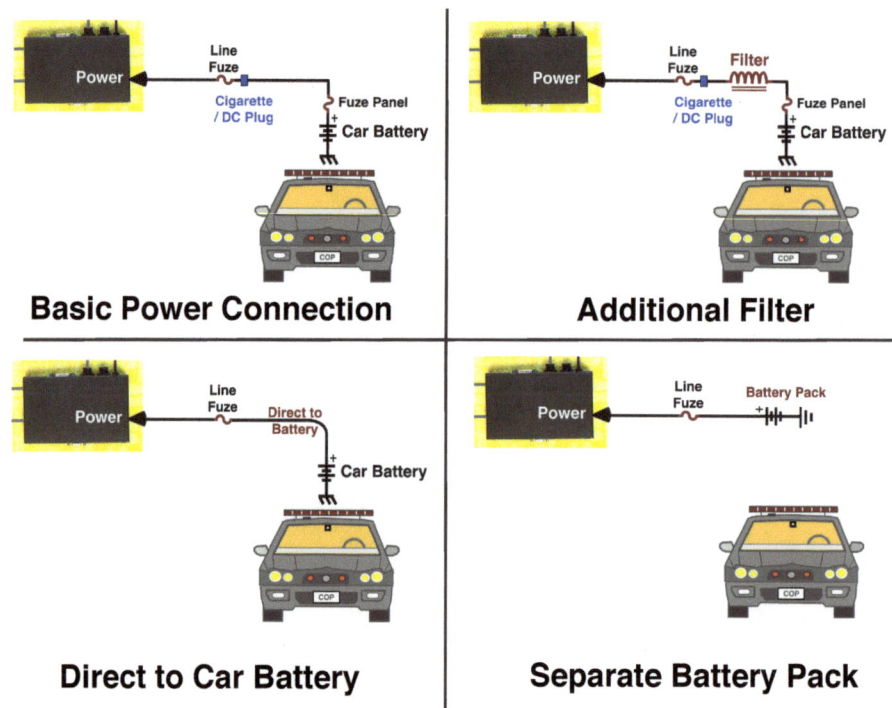

Basic Power Connection   **Additional Filter**

Direct to Car Battery   **Separate Battery Pack**

1.6 - Paperwork Traceability

All police radar's function about the same, but controls, status indicators, test procedures, and setting options are all different. It is important anyone operating a radar read, understand, and have a copy of the **radar manual**.

All radars and their tuning forks come with a **factory Certificate of Calibration**. Sometimes the factory certifications are in a single document, sometimes separate documents.

FACTORY DOCUMENTS
- **Radar Operator's Manual**
- **Radar Certificate of Calibration**
- **Tuning Forks Certificate of Calibration**

Maintenance records should be up to date and accurate.

- **Radar Maintenance Records**;
  - Repair Orders
  - Installation Date and Vehicle*
  - Speedometer Synchronization*
  - Radar / Dash-Camera Interface*
    *moving mode option.

Radar and tuning fork calibration records are important for documenting functionality and accuracy, and for court proceedings.

POLICE RECORDS
- **Radar Calibration Records**
- **Tuning Fork Calibration Records**
- **Event Logs / Officer Notes**

Most states require the radar and tuning forks be calibrated every 6 months.

Factory and Calibration Records

The factory paperwork should match the **radar**, and all radar **calibration records**. Information important to track includes the radar and tuning fork **serial numbers**, and the radar transmit **frequency**.

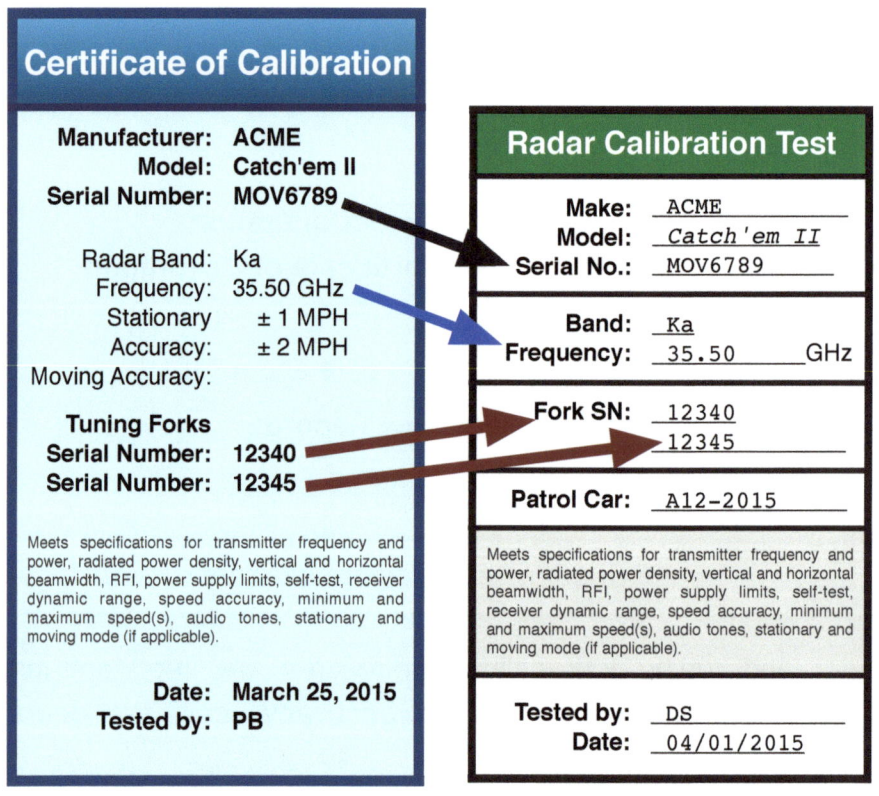

Typical Radar FACTORY
Certificate of Calibration

Typical Radar
Calibration *Record*

The radar and tuning fork calibration records should be up to date, and include corresponding serial numbers and radar operating frequency. Records for mounted radars should include patrol car identification, especially if the radar uses (connected to) the patrol car speedometer.

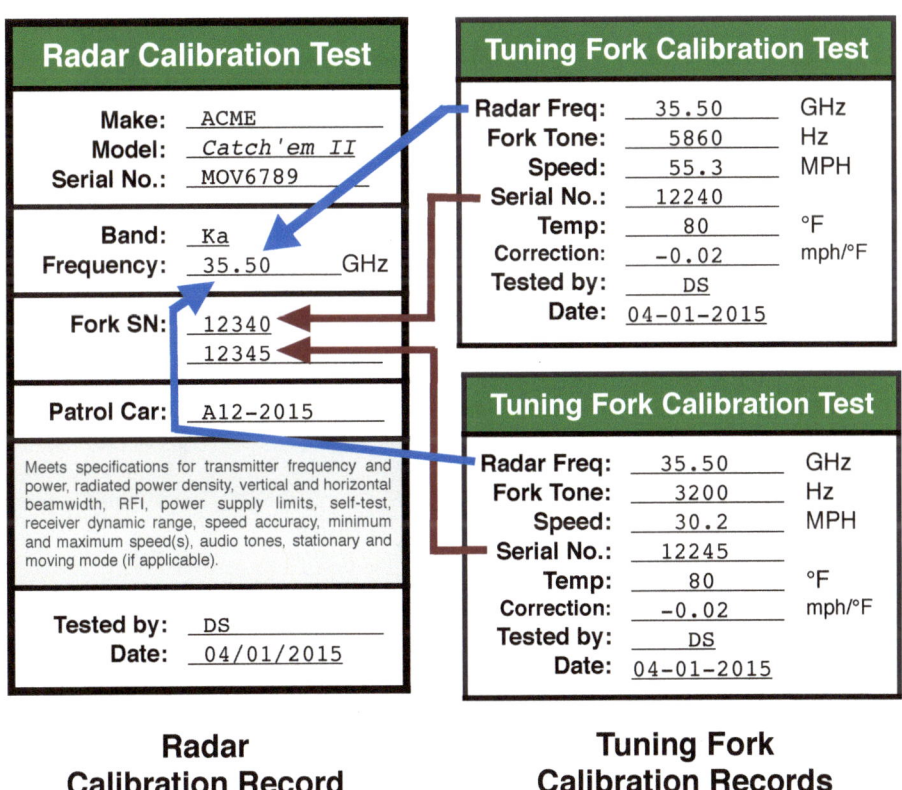

Radar Calibration Record

Tuning Fork Calibration Records

Calculate Tuning Fork Speed

For calculations:

Speed in MPH equals 0.3353 multiplied by Fork Resonance Tone in Hertz divided by radar frequency in GHz.

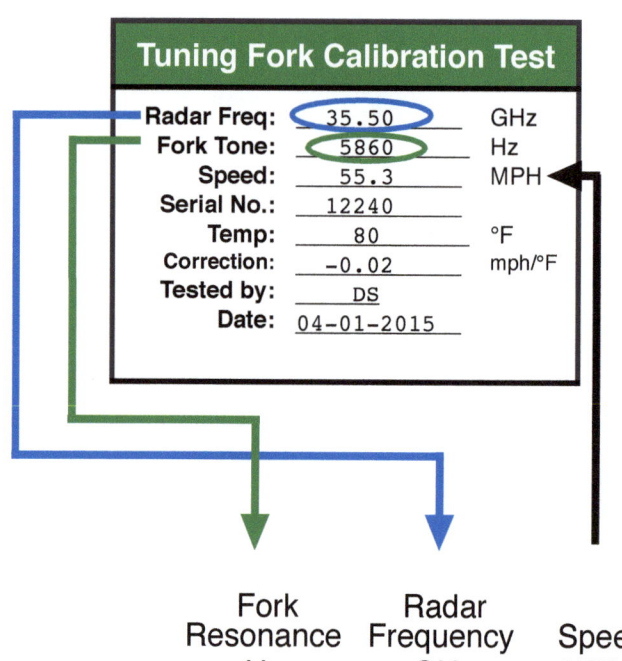

e.g.

		Fork Resonance Hz		Radar Frequency GHz	Speed **MPH**
0.3353 x		5860	/	35.50	= **55.3**

In Kilometers per Hour (KPH)

e.g.

		Fork Resonance Hz		Radar Frequency GHz	Speed **KPH**
0.5396 x		5860	/	35.50	= **89.1**

14

Chapter 2 - Tuning Fork Test

Many professionals involved with police radar do not understand the tuning fork test. Some operators do not do the test correctly, or do not do the test at all (and don't remember the procedure). This chapter details the proper procedures and protocols for testing radar speed accuracy using tuning forks.

A radar can measure a vibrating tuning fork and display a speed proportional to the fork resonant frequency -- the **tone**. The fork tone equals the *Radar Doppler* frequency speed. The speed depends on the radar **transmit frequency** and the **fork tone**.

The tuning fork is obviously not traveling at the speed the radar registers, the radar is tricked. However, this is a good trick and useful indicator.

Tuning fork test should be conducted daily, at start and end of shift at a minimum. Additionally, the test should be conducted in an area with no traffic or moving objects, especially in the beam. The antenna should be pointed up into the sky if possible.

Radar Modes Tested Using Tuning Forks.

- 1.) **STATIONARY** mode
- 2.) **MOVING** mode **Opposite Direction** traffic
- 3.) **MOVING** mode **Same - Direction** traffic

Each **mode** displays *different speeds* using the same set of tuning forks.

Front and rear antennas MUST be tested separately.

15

2.1 - Tuning Forks

Stationary only radars come with at least one tuning fork, tuned for that radar.

• The tuning fork **tone** simulates a **speed**.

• The *higher* the tone, the *higher* the speed.

• Tuning forks are tuned to a **specific radar**.

Tuning Fork with a
5800 Hertz (Hz) Tone

All tuning forks are labeled with **SPEED** in **MPH** (or **KPH**), and **Serial Number**. Many labels also include **radar band** (X, **K**, or **Ka**) and/or **Radar Transmit Frequency (GHz)**.

Speed:	**55 MPH -->**
Radar Band:	**KA BAND -->**
Radar Frequency:	**35.50 GHz -->**

Serial Number: -->

Typical Tuning Fork Identifications

Speed, Fork Tone (Doppler), and Transmit Frequency Conversions

$$V = 0.3353 \left(f_r / f_x \right) \qquad f_r = 2.9823 \left(V \, f_x \right) \qquad f_x = 0.3353 \left(f_r / V \right)$$

V = Speed (**MPH**)
f_r = **Fork** Tone (**Hz**)
f_x = Radar **Transmit** Frequency (**GHz**)

2.2 - Stationary Mode Test

POWER ON and MANUAL SELF-TEST

Self-test runs on power-up, or when the operator initiates it. The radar first illuminates all lights; mode indicators, speeds, status,

etc. The *operator must observe* if any lights or segments are burned-out, if so remove radar from service.

Next, the radar will display one of more of the following;

- **Range** or **Sensitivity** Setting (set by operator)
- **Audio** Volume (set by operator)
- Minimum Patrol Speed (set by operator)
- Internal Test Speeds
- Software Version
- Frequency
- Radar Operating Temperature
- Patrol Car Battery Voltage

Some information is in code, to fit on the display.

Self test only takes seconds, and will display a *PASS* or *FAIL* code, some radars also use audio beeps and buzzes. A

Passed Self-Test

radar that fails also displays a fault code. Some codes are listed in the manual, some codes only the factory can interrupt (if lucky). A fault will disable the radar until corrected.

RADAR SETTINGS FOR TUNING FORK TEST

1.)   Set **Range** or **Sensitivity** Setting to **Maximum**.

2.)   Select **Stationary** mode.

3.)   Select **Antenna** for test -- Front or Rear (if any).

4.)   Radar **Fan Filter OFF** (if any - read the manual).

5.)   Put in **TEST** mode if required (read the manual).

Test steps:

• Strike the fork **narrow edge** near the top on a hard **non-metallic surface**, such as wood or plastic.   Striking a fork on metal will detune the fork over time.

• Place the **narrow edge** of the fork about an **inch from the front** of the antenna.

• The radar should register the speed associated with the tuning fork as the target speed, within radar specifications. Stationary radar accuracy is typically plus or minus (±) 1 MPH.

• The radar **audio tone** should be clean and clear (not noisy).

• The test should be repeated with the second tuning fork for radars supplied with 2 forks (all moving mode radars).

Front and rear antennas MUST be tested separately.

18

2.3 - Moving Mode Test

Even if a moving mode radar will only be used from a stationary position, moving mode tuning fork tests must be performed to insure the radar is fully functional. A radar that cannot pass the moving mode tuning fork test could have accuracy problems in stationary mode.

All moving mode radars are supplied with 2 tuning forks that have different tones (speeds). One tuning fork simulates the patrol car speed, the other fork simulates target speed.

Typical Tuning Fork Set
Penny not included

The above tuning forks are tuned for a radar that transmits at 35.50 GHz. The smaller tuning fork has a 5800 Hertz tone and simulates a speed of 55 MPH. The larger tuning fork has a 3200 Hertz tone and simulates a speed of 30 MPH.

Both forks are used simultaneously to test moving mode. The order the forks are placed in front of the antenna and speed readings are different depending on radar mode.

Moving Mode OPPOSITE DIRECTION Traffic

Put radar in: **Moving mode**
    select: **Opposite Direction** traffic

TEST STEPS

1 *Strike* the **SLOWER** (Larger) fork, then the **Faster** (smaller) fork. Order does not matter, but larger slower fork oscillates longer.	e.g. **30 MPH** e.g. **55 MPH**
2 Place the **SLOWER** fork in front of antenna FIRST, then the **Faster** fork.	***30 MPH*** ***55 MPH***
3 The **SLOWER** fork should be the **PATROL** Speed*	**30 MPH**
DIFFERENCE in forks should be **TARGET** Speed*	**25 MPH**
3a Radar audio **TARGET TONE** should be clear.	**Good Tone**

* Within specifications, typically ± 1 MPH Target Speed
± 2 MPH Patrol Speed

TARGET   **PATROL**

Fast-Slow Fork   Slow Fork
55 - 30 = 25   30 MPH

Moving Mode Opposite Direction **Traffic**

Front and rear antennas MUST be tested separately.

Moving Mode SAME-LANE Traffic

Still in Moving mode,
change to **Same-Direction** traffic.

TEST STEPS

1 *Strike* the **SLOWER** (Larger) fork,	e.g. **30 MPH**
then the **Faster** (smaller) fork.	e.g. **55 MPH**
Order does not matter, but larger slower fork oscillates longer.	
2 *THIS TIME*	
Place the **Faster** fork in front of antenna FIRST,	*55 MPH*
then the **SLOWER** fork.	*30 MPH*
3 The **Faster** fork should be the **PATROL** Speed*	**55 MPH**
The *SUM* of the forks should be **TARGET** Speed*	**85 MPH**
3a Radar audio **TARGET TONE** should be clear.	**Good Tone**

* Within specifications, typically ± 1 MPH Target Speed
± 2 MPH Patrol Speed

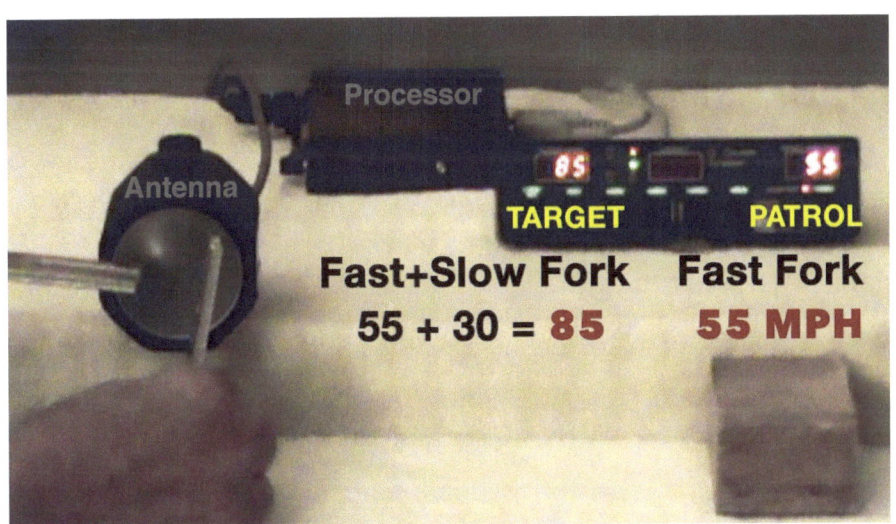

Moving Mode Same-Direction Traffic

Front   and rear

antennas MUST be tested separately.

Chapter 3 - How It Works

3.1 - Detection Range

Radar maximum detection range varies with radar and radar mode, target vehicle size and shape, weather, and electromagnetic noise pollution.

Maximum detection range for **stationary** mode and **moving mode opposite direction** traffic can be as low as **100 feet** or less to over **a mile**. Usually the larger the target vehicle the greater the radar detection range, but not always.

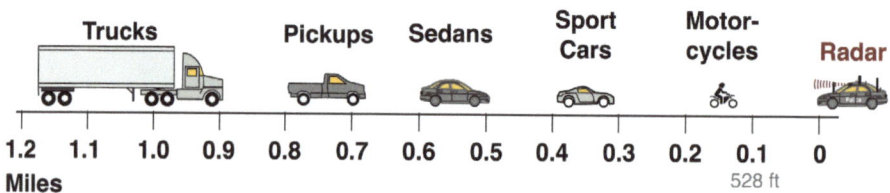

Moving mode **same-direction traffic** has shorter detection ranges. Same-direction vehicles traveling *slower* than the patrol car have even shorter detection ranges than vehicles traveling *faster*.

Weather inhibits radar operation by blocking the transmit and reflection signals.   Rain, sleet, snow, and fog can render a radar's detection range slightly to completely useless.   Strong **winds** can carry anything from dust to large debris that will cause problems.

Electrical and electromagnetic **interference** can reduce detection range, completely mask targets, and/or induce false speeds.   Common interference sources include other police radars, high voltage lines and transformers, and near-by transmitters (including in the patrol car and near-by patrol cars). See Chapter 5.4 - Electromagnetic Interference, pages 38-39.

3.2 - Beamwidth

Beamwidths vary with model from **9°** to over **15°**.   Vehicles outside the beam will not be processed or detected.

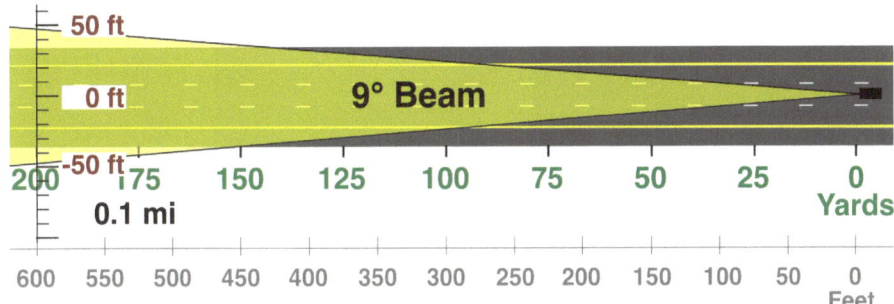

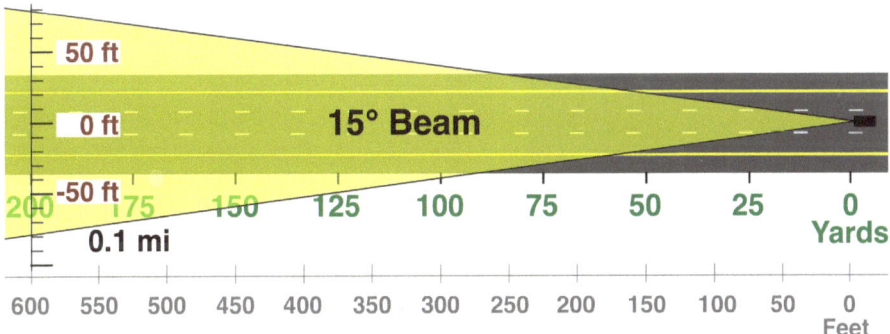

The equation to calculate beam spread for a particular range is 2 multiplied by Range multiplied by the tangent of half the beamwidth;

Beam Spread = 2 x Range x *tan* (Beamwidth / 2)

Beam Spread and *Range* in the same units (feet, meters, etc.).

23

3.3 - The Cosine Effect

The Cosine Effect causes target speed to measure low. It also causes target speed to change, the closer the target the faster the speed changes. When speed changes too fast the radar cannot process the reflection and loses track (*Blind Zone*). The closer a radar is to the road, the smaller the cosine angle and the smaller the blind zone.

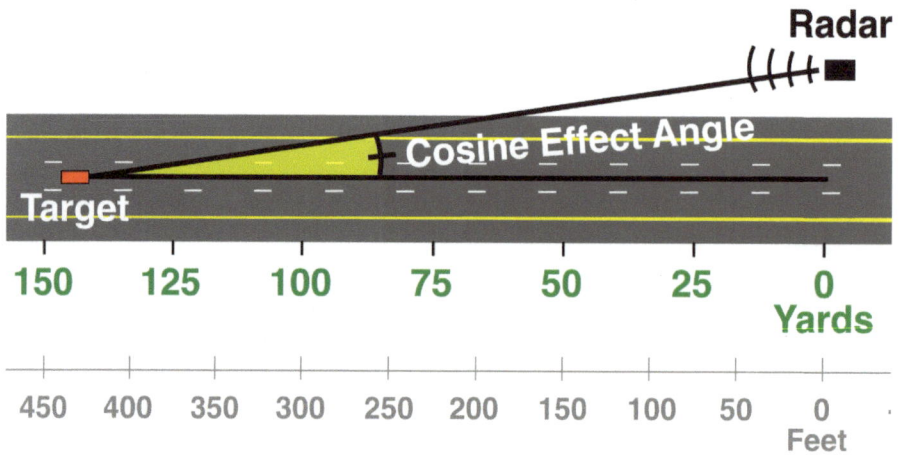

Cosine Effect Angle

When target is even with the radar the Cosine Speed is 0 MPH.

Cosine of 90° is 0.

Cosine Effect SPEED

Measured Speed = True Speed x *cos* (Angle)

$$V_m = V \, cos \, (B)$$

V_m = Radar Measured Speed

V   = Target Vehicle True Speed

B   = Cosine Effect Angle, *changes with range*

24

Radar Blind Zone

The blind zone is a function of;

• Radar Distance from Target Vehicle Lane,
• Target Speed,
- Radar Accuracy,
- Radar Sample Time.

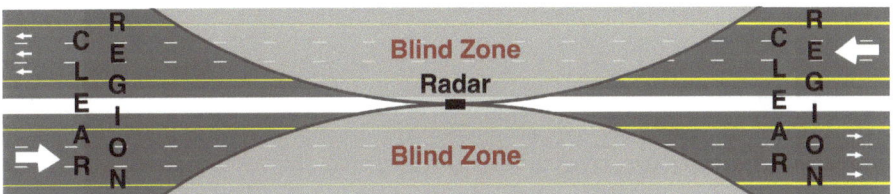

Radar Clear Region and Blind Zone

The *faster* the target and/or the *greater the radar distance* from the target lane, the larger the blind zone. The blind zone applies to on-coming or receding traffic. It also applies to moving mode opposite direction traffic, expect the blind zone is about double that of stationary mode. The blind zone is NOT a function of the radar beam or alignment.

Chapter 4 - Operation

4.1 - Daily Test and Use

RADAR TEST BEFORE USE
• Make sure the Radar is properly **Calibrated**. -- Make sure **tuning fork**(s) calibrated (that belong with the radar).
• Run Radar **Self-Test**. -- Power-on Self-test, then run **manual** self-test.
The Following Test Should be Done on Both Antennas. • **Tuning Fork Test**.* -- Test at start and end of shift, minimum. -- Moving mode requires 2 tuning forks. -- Some radars have additional settings for this test (read the manual).
• Test drive radar **patrol speed** matches **speedometer**.*
• Check using a controlled **Test Vehicle**. -- Test vehicle should have a calibrated speedometer. Not required by manufacturers, but most recommend.

*A moving mode radar that cannot pass the moving mode tuning fork test or cannot match radar measured patrol speed with the speedometer, could have accuracy problems in stationary mode.

DAILY USE

• Target must be traveling at a relatively **constant speed**.

• Target vehicle in the radar **beam**.
  This one seems obvious, but it is documented NOT to always be the case.

• Radar measures strongest reflection,
  usually closest vehicle, but not always.

4.2 - Track History

Radars use sensitive microwave receivers subject to false readings. Some false readings are short, momentary to a few seconds, some are persistent. It is critical to establish a good **TRACK HISTORY - a steady stable reading** and **clear audio tone** for at least 3 seconds in stationary mode, longer in moving mode or adverse situations.

Obvious Bogus Readings

4.3 - Stationary Mode

Microwave (and laser) radar is accurate only when the target speed is relatively constant and traveling on a **straight** and **flat** road, with the radar positioned **close to the targeted vehicle lane**.

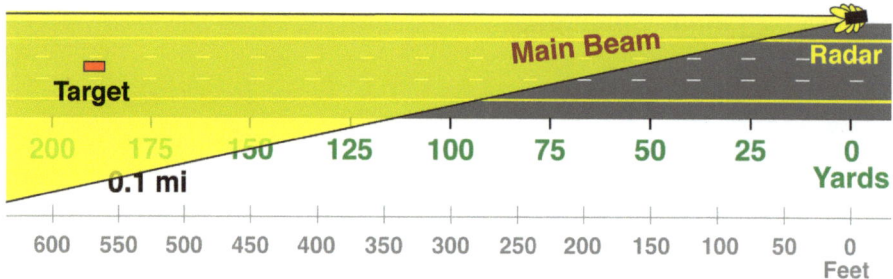

Optimum Setup - *Close to Road*, 12° Beam Aligned 6° to Road

RADAR SETUP

- Radar should be **close to the road**,
  to minimize *Cosine Effect* limitations and Blind Zones.

- Road **straight** and **flat**.

- Set **Fan** and/or **Interference Filters** if necessary.

- Adjust **Radar Range / Sensitivity Setting** for conditions.

Improper Setup

A radar aimed at anything near 90° to the road, and traffic, will **not measure any speeds**. The radar antenna angle to traffic needs to be at or close to 0°. The larger the angle the larger the blind zone (too close). **Targets must be *COMING AT* the radar,** or as close to *coming at* the radar as possible.

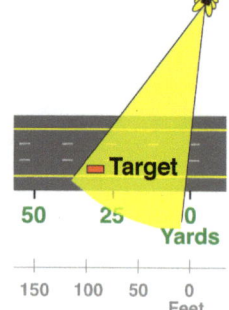

28

Radar Performance for Different Locations

The following examples account for the *Cosine Effect* only, and assumes the target is in the beam. Speed drops faster as target gets closer. Radar loses track when speed drops too fast.

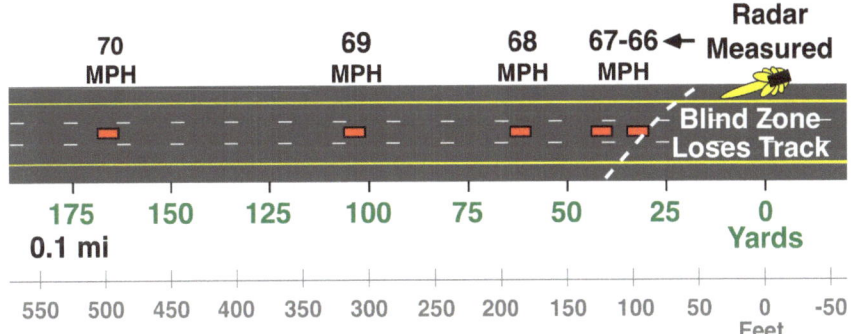

Target Speed = 70 MPH
Radar 10 Feet from Highway

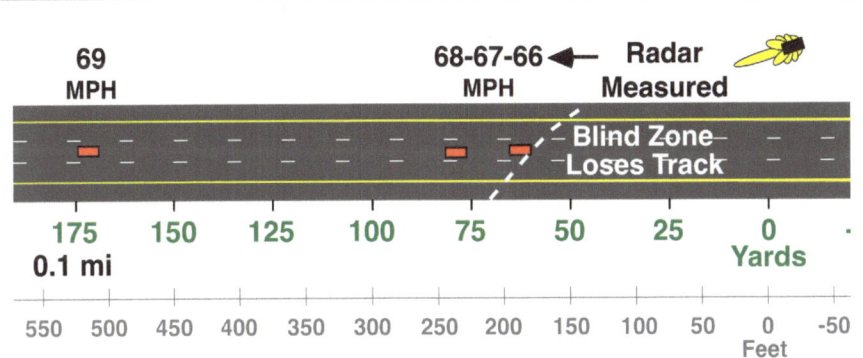

Target Speed = 70 MPH
Radar 50 Feet from Highway

A radar 50 feet off the highway will measure a 70 MPH vehicle at 70 MPH around 200 yards *or greater*. Also, the Blind Zone is larger compared to a radar 10 feet off the road.

4.4 - Moving Mode Opposite Direction Traffic

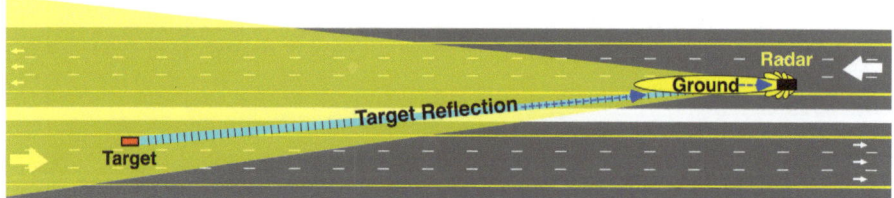

Scenario

MOVING MODE USE
in addition to stationary mode considerations

• Check radar **measured patrol speed** matches **speedometer during traffic measurements**.
A radar connected to the speedometer (VSS) automatically checks speeds.

• **No traffic between targeted vehicle and radar.**
Same lane traffic can be mistakenly used as the ground echo causing speed errors (*Patrol Speed Shadowing from moving vehicles*).

• **Target** and **Patrol** vehicles at relatively **constant speed**.

• Patrol car must be moving at radar **minimum speed.**
Varies with radar model, typically 5 - 20 MPH.

• **Blind zone** about double that of stationary mode.

On-coming police cruiser doing 44 MPH in a 35 MPH zone
Radar vehicle at 23 MPH (matched speedometer) and *slowly* accelerating

4.5 - Moving Mode Same-Direction Traffic

Additional Restrictions for Moving Mode Same-Lane
in addition to all other mode considerations and restrictions.

- Target at least **3** to **5 MPH** *faster* or *slower* than patrol car.
Faster preferred because radar has better detection.

- **No moving vehicles** between **target** vehicle and **patrol** car.
Includes opposite direction traffic for slow patrol speeds or close on-coming lanes.

Same-lane vehicle at 47 MPH in a 35 MPH zone
Radar vehicle at 32 MPH (matched speedometer)

Cosine Effect angle is 0° for target vehicle's coming directly at or away from the radar antenna. No Cosine error here.

4.6 - Fastest Target Mode

Many radars can measure 2 target vehicles simultaneously, the **strongest** reflection and the **fastest**. If the fastest vehicle is also the strongest reflection, only this target's speed is displayed. The fastest option is available in all radar modes. Also see Chapter 1.2 - Radar Displays, page 4.

Chapter 5 - Limitations

• Establish a valid **Track History** (Chapter 4,2, page 27).

• **Cosine Effect angle MUST be small**, radar must be close to traffic lane (Chapter 3.3, page 24).

• Moving mode radar **antennas** must be mounted to the patrol vehicle **STRAIGHT** and **LEVEL** (Chapter 1.3 - Antenna Alignment page 5-7) to avoid *Patrol Speed Shadowing* (pages 41 - 42) errors.

5.1 - Radar Accuracy

Police radars are accurate to plus or minus (±) 1 MPH in stationary mode, and ± 2 MPH in moving modes. Radar measured patrol speed is accurate to ± 1 MPH. Some radars claim ± 0.5 MPH accuracy for stationary mode, ± 1 MPH for moving mode.

Typical Radar Accuracy

Stationary mode target: ± 1 MPH (or ± 0.5 MPH)

Moving mode target: ± 2 MPH (or ± 1.0 MPH)
Patrol vehicle: ± 1 MPH (or ± 0.5 MPH)

Target and Patrol Acceleration Limits

Police radars are limited to measuring vehicles traveling at a relatively constant speed. Vehicles changing speed greater than radar accuracy during one sample period (minimum time to get one reading) cannot be measured, speed is changing too fast.

Acceleration Limit = Accuracy / Sample Time

± 1.0 MPH Accuracy   **± 0.5 MPH Accuracy**

Sample Time	Maximum Speed Change	Sample Time	Maximum Speed Change
0.2 sec	± 5.0 MPH / sec	**0.2 sec**	**± 2.5 MPH / sec**
0.3 sec	**± 3.3 MPH / sec**	0.3 sec	± 1.7 MPH / sec
0.4 sec	± 2.5 MPH / sec	0.4 sec	± 1.3 MPH / sec
0.5 sec	± 2.0 MPH / sec	0.5 sec	± 1.0 MPH / sec

A radar's acceleration limit is based on it's *Sample Time* and *Stationary Mode Accuracy*.

In general, vehicles changing speed greater than around ± **3 MPH per second cannot be measured** by most radars.

5.2 - Target Vehicle Selectivity

Multiple close vehicles make it difficult or impossible for the operator to know which vehicle the radar is tracking. In steady moving dense traffic a different vehicle is displayed every fraction or so of a second. In these cases a reliable track history is impossible to establish. Also, the displayed speed is not necessarily the closest. The closest vehicle may be in the radar Blind Zone, especially when the radar is far from the traffic lanes.

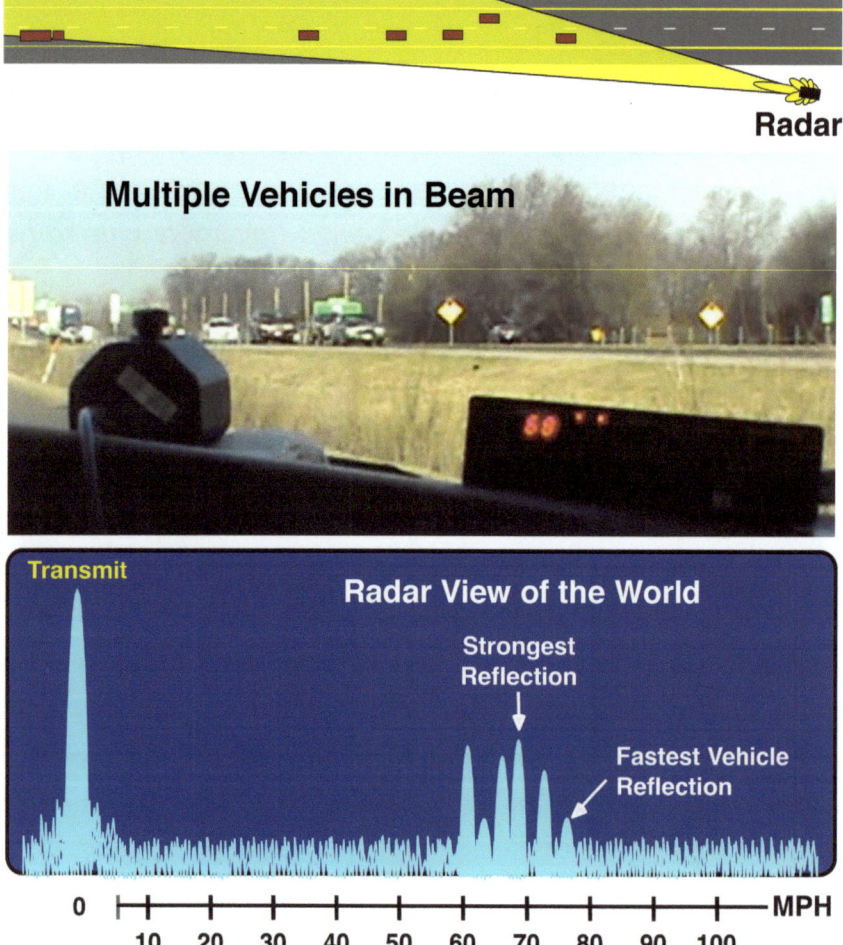

5.3 - Multi-path Errors

Multi path occurs when the transmit signal does not reflect directly from the radar to target and back, but from radar to a moving target to another moving target(s) than back to the radar. The more moving targets relatively close to each other, the greater the chance for multi-path errors. Multi-path errors can be a few MPH to tens of MPH, and can be momentary to several seconds long.

Multi-path Error

5.4 - Electromagnetic Interference

Police radars should only be used in areas relatively free of electromagnetic sources such as **transmitters** and **high voltage apparatus**.

A few Electromagnetic Interference Sources

Police or **Safety Radars** in the area

Field Disturbance Transmitter Sensors
Shares Frequencies with Police Radars.
- Traffic Signals
- Door Openers
- Burglar Alarms
- Obstruction Detectors
  Railroad Rolling Stock and Locomotives
  Moving Farm Equipment
  Fork Lifts

Neon Signs

Electrical Power
- Power Lines and Pole Transformers
- Generator Sub-stations
- Transformer Sub-stations
- High Voltage Transmissions Lines

Communication Towers
- TV / AM / FM / Broadcast
- Cellular Phones
- WiFi Data
- Police / Fire Stations
- Business Radios
- Amateur Radios
- Microwave / Radio Relays

Hospitals

Airports
- Weather / Tracking Radar
- Communications
- Beacons

Military Installations

Weather Radar

Mutual Interference from Another Police Radar

Stationary radar measured target at 16 MPH, but target (police car) going much faster, probably close to the 70 MPH speed limit. The radar was completely desensitize after the interference and missed a number of vehicles until the radar re-adjusted receiver gain.

Police vehicles have multiple transmitters **known** to have the potential to interfere with the radar. Radios, walkie-talkies, and personal communication devices, in and near the patrol vehicle, should not be used during a radar measurement.

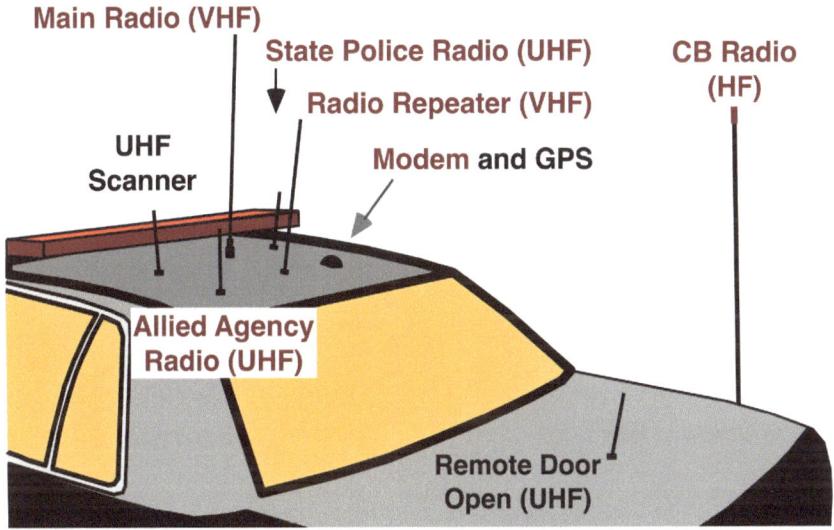

Main Radio (VHF)

State Police Radio (UHF)

CB Radio (HF)

Radio Repeater (VHF)

UHF Scanner

Modem and GPS

Allied Agency Radio (UHF)

Remote Door Open (UHF)

Patrol Car Antennas

Interference Effects

- **Reduce Detection Range***
Automatic gain adjust out interference and detection range.

- **Produce False Speed Readings***

- **Completely Mask Legitimate Targets***

- **Activate Radio Frequency Interference (RFI) mode**
Stops processing speeds when interference detected (not reliable*).

* No radar indication to the operator.

Some interference will produce audio noise.

5.5 - Transmit Turn-On Time

Instant-on radars are not exactly instantaneous, some turn-on or warm-up time is required. It takes a minimum of about 0.3 seconds, but can take up to 2 seconds, to reach a steady stable transmit signal. Speed measurements, if any, during warm-up are completely unreliable.

One manufacturer has models with a pulsed mode, radar transmits a short 0.067 second burst that some radar detectors cannot pick up. The operator uses the pulsed mode to get an *estimate* of speed without setting off some radar detectors. The operator cannot lock any speeds but must switch to normal transmissions, and take another measurement. The burst is so short that speed measurements are **completely unreliable** and totally **useless,** and **counter productive** (waste of time).

5.6 - Patrol Speed Shadowing

Moving mode radar is designed to use a ground reflection from directly in front of the patrol car, to measure the patrol speed. Off angle ground reflections cause the patrol speed to measure **low** *(Patrol Speed Shadowing)*, and target speed to read **high**. A patrol speed low by 10 MPH, means target is high by 10 MPH. The error is due to the *Cosine Effect.*   Also see Chapter 1.3 - Antenna Alignment, pages 5 - 7.

Measured Patrol Speed = Actual Speed x *cos* (Angle Error)
The greater the speed and angle error, the greater the speed error.

The operator must monitor patrol speed against the speedometer to make sure there is not any speed shadowing. Radar's connected to the speedometer (VSS - velocity speed sensor) automatically detect shadowing and will re-establish a ground track until it matches the speedometer.

Common Patrol Speed Shadowing Sources

- Parked or Moving Vehicles
- Sound and Construction Barriers
- Guardrails and Cable Crash Barriers
- Highway Barrels (metal or water filled plastic)
- Overhead, Roadside, and Construction Signs
- Bridge Trusses / Underpasses / Pillars
- Hill Cuts / Depressions / Tunnels
- Plowed Snow Pile / Ice Patches / Ditch Water

Patrol Speed Shadowing

The **speedometer** measured 50 MPH, however the radar measured patrol speed at 20 MPH -- *low by 30 MPH.* This makes the radar measured **target** speed (79 MPH) *high by 30 MPH.* True target speed was 49 MPH (79 - 30).

	Radar Measured		ERROR Correction		True Speed
Patrol Car	20	+	30	=	*50*
Target Vehicle	79	-	30	=	*49*

Ground Angle = cos^{-1} (V_m / V)

V_m = Radar Measured Patrol Car Speed,
V = Speedometer Speed

In the above example the ground reflection was the inverse cosine of (20/50) = 66° off angle.

5.7 - Other Errors

Batching or Target Speed Bumping

Batching, better described as *Target Speed Bumping*, occurs in moving mode when the patrol speed suddenly changes causing a false or inaccurate reading. Radar measured target speed and patrol speed are not updated simultaneously, if patrol speed changes suddenly the radar may be using outdated data, leading to speed errors. Sudden acceleration, braking, turning, curves, or hitting bumps can cause batching or target speed bumping.

Scanning Error (intentional)

Scanning errors occur when a **hand held radar** is pointed or scanned in a direction that picks up the patrol car air-circulation fan or motor. Fan motion and motor electromagnetic noise can produce false speed readings. False readings are proportional to fan speed and reflection angle.

Panning Error (intentional)

Multi-unit radars will produce false readings when the antenna is panned around the display or processor. Panning the antenna at different distances and angles produces different false speeds.

Chapter 6 -- Theory

Police radars use the *Doppler Effect* to measure speed. The radar transmits a signal and receives reflections simultaneously.

1.) Radiates (transmits) a **microwave** frequency signal in a cone shaped **beam 9°** to **15°** wide.

2.) Reflections from moving objects have a slight **frequency difference** (*Doppler Effect*) from the transmit frequency.

3.) The frequency **difference** is based on object speed, and can be **calculated in MPH**.

6.1 - Radar Bands

Police radars in the United States transmit in either the **X**, **K**, or **Ka** band. The Ku band (13.45 GHz) is allocated for police radar use, but not used.

Band	Frequency	Notes
X	**10.525 GHz** Single Channel	• Relatively good performance in bad weather. • Vulnerable to *multiple* Interference sources. • Antenna diameter ≈ 4 inches. • Being phased out.
K	**24.150 GHz** Single Channel	• Range limited by rain, sleet, snow. • Vulnerable to Interference Sources. • Antenna diameter ≈ 3 inches
Ka	**33.4 - 36 GHz** Transmits on 1 of 13 Channels	• Range limited by rain, sleet, snow. • Small Antennas, diameter ≈ 2.5 inches. • Most popular.

1 GHz = 1 gigahertz = 1 Billion Hertz = 1,000,000,000 Hertz

Police Radar Frequency Bands

Police radars transmit at much higher frequencies than AM and FM radio, over-the-air TV, cellular phones, GPS satellites, or Wi-Fi.

Frequency Spectrum

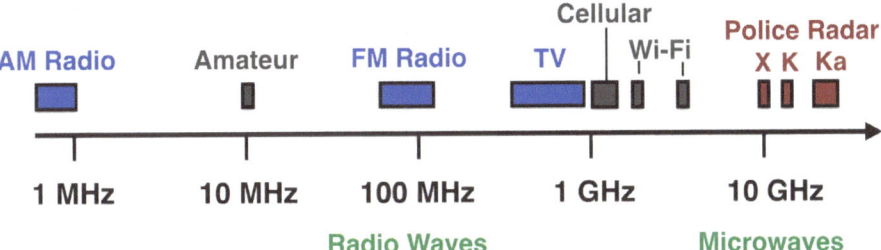

Frequencies	Band Designation	Services
300 - 3000 kHz	MF (Medium Frequency)	AM Radio
3 - 30 MHz	HF (High Frequency)	Short Wave Radio
30 - 300 MHz	VHF (Very High Frequency)	FM Radio
300 - 1000 MHz	UHF (Ultra High Frequency)	TV
1 - 2 GHz	L Band	Cellular Phones / GPS
2 - 4 GHz	S Band	Wi-Fi
4 - 8 GHz	C Band	Wi-Fi
8 - 12 GHz	**X Band**	Police Radar
12 - 18 GHz	Ku Band	Satellite TV
18 - 27 GHz	**K Band**	Police Radar
27 - 40 GHz	**Ka Band**	Police Radar
40 - 300 GHz	mm (millimeter)	millimeter wavelengths

kilohertz (kHz) = thousand Hertz = 1,000 Hertz
megahertz (MHz) = million Hertz = 1,000,000 Hertz
gigahertz (GHz) = billion Hertz = 1,000,000,000 Hertz

6.2 - Transmissions and Reflections

• Most of the transmit signal misses the target vehicle.

• Most of the signal reflected off the vehicle goes in directions other than the radar antenna.

• Reflections are Doppler Shifted (frequency changes) for moving vehicles. The reflected frequency is higher for on-coming vehicles, and lower for receding vehicles.

• Radio and (radar) microwaves travel at the speed of light - 670,616,625 MPH.

• Microwaves are "Line-of-Sight", and do not bend around objects like lower frequency radio waves.

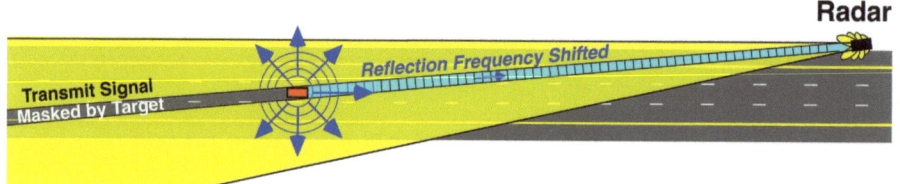

Scenario
Doppler Reflections

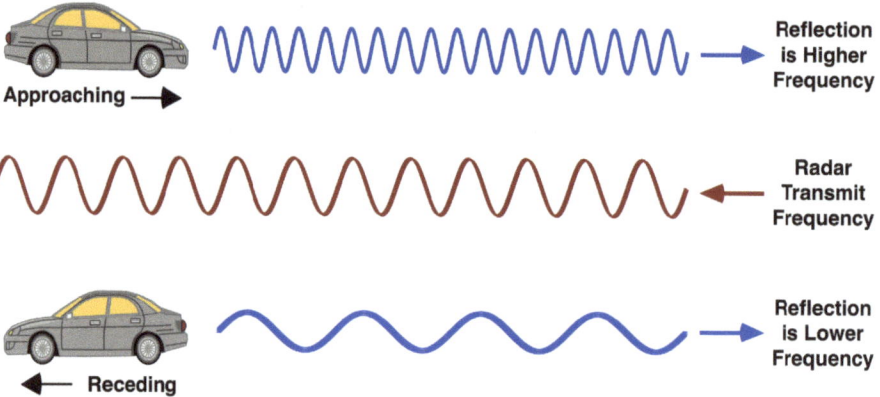

Stationary Radar Frequency Spectrum

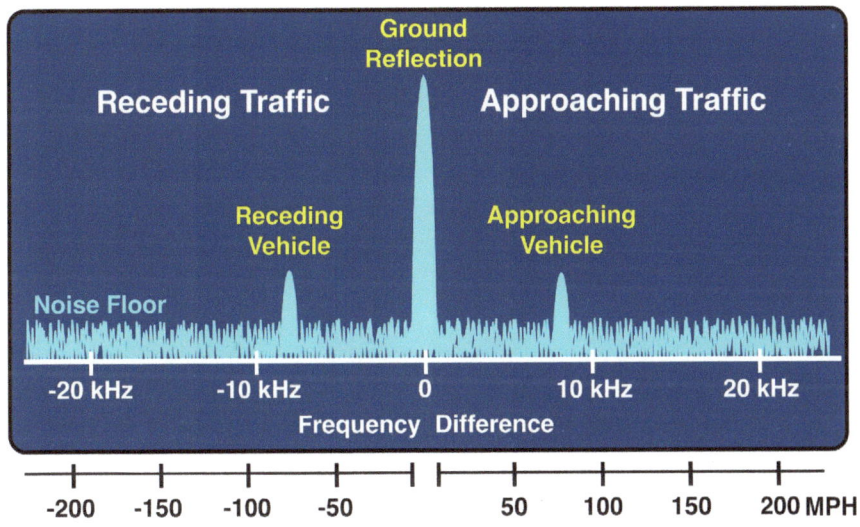

Radar View of the World

All stationary radars can measure approaching traffic, some also measure receding traffic (at the same time). Receding traffic has a lower audio tone for identifying direction, no visual indicator. Some models do give a visual indication of traffic direction; (+) for on-coming traffic, (-) for receding traffic. Some radars allow the operator to select a traffic direction.

Traffic Direction Variations (varies with model)

- Approaching traffic only

- Approaching and receding traffic,
  - only an audio indication of traffic direction

- Approaching and receding traffic,
  - audio and visual indication of traffic direction

- Operator selects traffic direction.

Moving Mode
Moving Mode Radar Frequency Spectrum

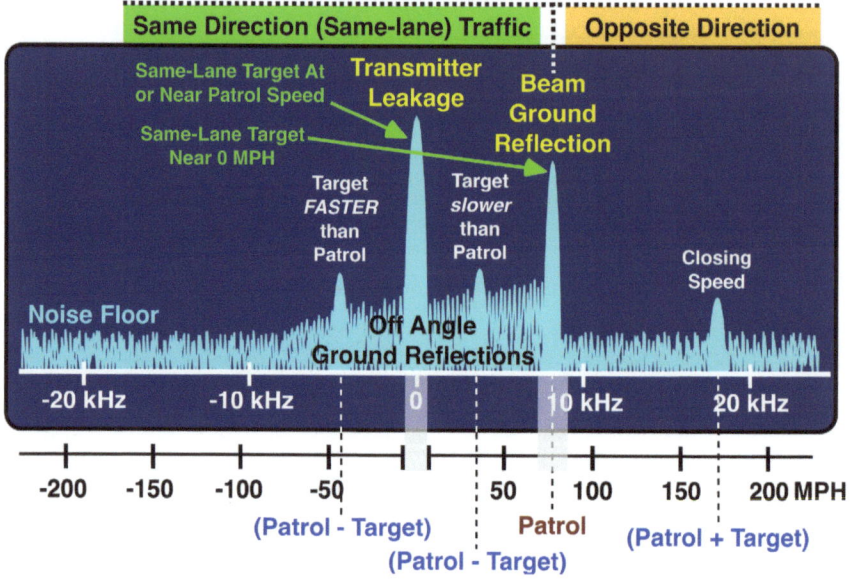

Radar View of the World in Moving Mode

In moving mode the operator selects either opposite direction traffic, or same-direction traffic.   Not all moving mode radars have same-direction capability.   Same direction traffic has a shorter detection range because the vehicle reflections compete with antenna side and back lobe ground reflections.   Target vehicles traveling slower than the patrol vehicle have the shortest detection range.

END

Radar versus Lidar
Microwave radar versus infrared laser radar (lidar)

	Radar	**Lidar**
Operation	Stationary or Moving	Stationary
Aim (pointing)	Easy Aim or Fixed Mounted	***Exact* Aim Required**
Traffic Conditions	Light to Moderate	Light to Dense
Range (distance)	Short to Long Range	Short Range
Measurements	Speed from Strongest Reflection. Some radars track 2 vehicles, strongest reflection & fastest.	Speed and Range
Transmissions	Seconds to Continuous	Seconds

Lidar (lasers) should **not** be operated from behind glass (windshield, etc.).

US Interstates
Typical Parameters

Lane Width:	12 feet
Left (inside) Shoulder Width	
Rural or Urban:	4 feet
3 or More Lanes Each Direction:	10 feet
Heavy Truck Traffic:	12 feet
Mountains:	4 feet
Mountains, 4 or More Lanes Each Direction:	8 feet
Right (outside) Shoulder Width	
Rural or Urban:	10 feet
Heavy Truck Traffic:	12 feet
Mountains:	8 feet
Median Width	
Rural:	36 feet
Mountains or Urban:	10 feet
Vertical Clearance	
Rural:	16 feet
Urban:	14 feet
Sign Supports / Pedestrian Overpasses:	17 feet

Standard Railroad Gauge (US)
4 feet, 8.5 inches

Police 10 Codes

Police and other agencies use "**10**" **codes** to verbally communicate efficiently. The word 10 indicates the next number (or numbers) is code. Four codes are widely used throughout the United States and listed below. APCO - Association of Police Communications Officers.

Code	General Purpose	APCO	Norfolk, VA	Walnut Creek, CA
10-0		Use Caution		
10-1	Unable to Copy - Change Location	Signal Weak	Police Officer Needs Help	Poor Radio Reception
10-2	Signal Good	Signal Good	Assist Officer	Good Radio Reception
10-3	**Stop Transmitting**	**Stop Transmitting**	**Clear the Air - Emergency**	**Stop Transmission**
10-4	Acknowledgment (OK)	**Affirmative (OK)**	Acknowledgment (OK)	**Message Received**
10-5	Relay	Relay To / From	See a Complainant	Relay Message
10-6	Busy- Unless Urgent	Busy	Investigation Police or Fire	Change Radio Channel
10-7	**Out of Service**	**Out of Service**	**Out of Service - Off Air Subject to Call**	**Out of Service**
10-8	**In Service**	**In Service**	**In Service**	**In Service**
10-9	Repeat	Say Again	Arrive at Scene	Repeat Message
10-10	Fight in Progress	Negative	Traffic Detail	Off Duty
10-11	Dog Case	__On Duty (Employee Number)	Broken Glass	Visitors Can Hear Radio
10-12	Stand By (Stop)	Stand By (Stop)	Vandalism	Advise Weather / Road Conditions

Code	General Purpose	APCO	Norfolk, VA	Walnut Creek, CA
10-13	Weather-Road Report	Weather Conditions	Leaking Water Main or Sewer Hole in Street / Sidewalk	
10-14	Prowler Report	Message / Information	Convoy or Escort	
10-15	Civil Disturbance	Message Delivered	Prisoner in Custody	Prisoner in Custody
10-16	Domestic Problem	Reply to Message	Pick Up Prisoner	Pick Up
10-17	Meet Complainant	En-route	Administrative Assistance	Getting Fuel
10-18	Quickly	Urgent	Detail	
10-19	Return to ___	(In) Contact	Return to Station	Return or Go to ___
10-20	**Location**	**Location**	**What is Your Location**	**Location**
10-21	**Call (__) by Phone**	**Call (__) by Phone**	**Call (__) by Phone**	**Telephone**
10-22	Disregard	Disregard	Investigate a Break-In	Cancel or Disregard
10-23	Arrived at Scene	Arrived at Scene	Breaking-In (In Progress)	Stand-By
10-24	Assignment Completed	Assignment Completed	Someone in the Building	
10-25	Report in Person (Meet)	Report To (Meet)	Prowler	Do You Have Contact With ___?
10-26	Detaining Subject, Expedite	Estimated Arrival Time (ETA)	Larceny	Clear of Warrants
10-27	(Driver) License Information	License / Permit Information	Rape Report	Subject Wanted
10-28	**Vehicle Registration Information**	**Vehicle Information**	**Check FULL Registration, License, Motor, Name, Stolen**	**Registration Check**
10-29	Check for Wanted	Records Check	Person with a Gun	Check for Warrants

Code	General Purpose	APCO	Norfolk, VA	Walnut Creek, CA
10-30	Unnecessary Use of Radio	Danger / Caution	(a) Vehicle Accident (b) Vehicle Accident Personal Injury (c) Hit and Run	
10-31	Crime in Progress	Pice Up	Hold Up and Robbery	
10-32	Man with Gun	__ Units Needed (Specify)	Defective Traffic Light	
10-33	Emergency	Need Immediate Assistance	Execute Warrant	Alarm is Sounding
10-34	Riot	Current Time	Narcotics Investigation	
10-35	Major Crime Alert		Get a Stolen Auto Report	Time Check
10-36	**Correct Time**		**Correct Time**	**Correct Time**
10-37	(Investigate) Suspicious Vehicle		Finished with Last Assignment	Please Identify Your Unit
10-38	Stopping Suspicious Vehicle		(a) Reckless Driving (b) Drunk Driving	
10-39	Urgent-Use Light, Siren		Report of a Dead Person	Can __ Come to Radio?
10-40	Silent Run- No Light, Siren	Fight in Progress	Suspicious Person-Auto	Is __ Available for Phone Call?
10-41	Beginning Tour of Duty	Beginning Tour of Duty	(a) Lost Child (b) Investigate Runaway	
10-42	Ending Tour of Duty	Ending Tour of Duty	Car Improperly Parked	
10-43	Information	In Pursuit	Drunk	
10-44	Permission to Leave __ for __	Riot	Disturbance (type)	
10-45	Animal Carcass at __	Bomb Threat	Fight	Subject Condition: A to D

Code	General Purpose	APCO	Norfolk, VA	Walnut Creek, CA
10-46	Assist Motorist	Bank Alarm	Attempt Suicide	
10-47	Emergency Road Repair at __	Complete Assignment Quickly	(a) Injured (b) Sick (c) Demented Person (d) Maternity Case, State, Civilian, Fire, or Police	
10-48	Traffic Standard Repair at __	Detaining Suspect, Expedite	Person Overboard	
10-49	Traffic Light Out at __	Drag Racing	Braking Dog	Proceeding to __
10-50	Accident (F,PI,PD)	Vehicle Accident, PD,PI, F	Court Cases	Drugged
10-51	Wrecker Needed	Dispatch Wrecker	General Message	Drunk
10-52	Ambulance Needed	Dispatch Ambulance	Open Door/ Window (State Which)	Ambulance Needed
10-53	Road Blocked at __	Road Blocked	Gas-Repairs- Wash	Person Down
10-54	Livestock on Highway	Hit and Run Accident, PD, PI, F	Man Molesting Children	Possible Body
10-55	Intoxicated Driver	Intoxicated Driver	Bomb Threat	Coroner's Case
10-56	Intoxicated Pedestrian	Intoxicated Pedestrian	Unruly Crowd	Suicide (a) Attempted
10-57	Hit and Run (F, PI, PD)	Request BT Operator	Tampering With Automobile	
10-58	Direct Traffic	Direct Traffic	Burglar Alarm	
10-59	Convoy or Escort	Escort	Traffic Violator	Security Check

Code	General Purpose	APCO	Norfolk, VA	Walnut Creek, CA
10-60	Squad in Vicinity	Suspicious Vehicle	(a) Dead Dog (b) Live Dog (c) Female or Stray (d) Dog Bite	
10-61	Personnel in Area	Stopping Suspicious Vehicle	Void IBM Card	Bike Theft
10-62	Reply to Message	Breaking and Entering (B & E) in Progress	Radio Test	
10-63	Prepare to Make Written Copy	Prepare to Receive Assignment	Personal Relief	Prepare to Copy
10-64	Message for Local Delivery	Crime in Progress	Eating (State Location)	
10-65	Net Message Assignment	Armed Robbery	Exposure	
10-66	Message Cancellation	Notify Medical Examiner	Send Wrecker to (a) Owner Request (b) Police Request	Suspicious Person
10-67	Clear for Net Message	Report of Death	Smoke & Flames Visible	Person Calling for Help
10-68	Dispatch Information	Livestock in Roadway	In Commission on Stand-By	
10-69	Message Received	Advise Telephone Number	Held Up By Bridge or Train	
10-70	Fire Alarm	Improper Parked Vehicle	Danger / Caution	Prowler
10-71	Advise Nature of Fire	Improper Use of Radio	False Alarm	Shots Fired
10-72	Report progress on Fire	Prisoner in Custody	Person Found in Burning Building	
10-73	Smoke Report	Mental Subject	Existing Conditions	How Do You Copy?
10-74	Negative	Prison / Jail Break	En-route	
10-75	In Contact with __	Wanted or Stolen	Dispatch Mechanic	

Code	General Purpose	APCO	Norfolk, VA	Walnut Creek, CA
10-76	En Route __	Prowler	Rewind Box (Give Location)	
10-77	ETA (Estimated Time of Arrival)	Direct Traffic at Fire Scene	Send VEPCO (State Gas or Electric) (b) Send C and P	
10-78	Need Assistance		Held Up by (state)	
10-79	Notify Coroner		Courtesy Call	
10-80	Chase in Progress	Fire Alarm	Critical Call (Code Red)	Explosion
10-81	Breatherlizer Report	Nature of Fire	Alarm of Fire	
10-82	Reserve Lodging	Fire in Progress	Additional Engine Co.	
10-83	Work School Crossing at __	Smoke Visible	Additional Ladder Co.	
10-84	If Meeting __ Advise ETA	No Smoke Visible	Second Alarm	
10-85	Delay Due to __	Respond without Blue Lights / Siren	Third Alarm	
10-86	Officer / Operator on Duty		Person Trapped	Any Traffic for Me?
10-87	Pickup / Distribute Checks		Auto Fire	
10-88	Present Telephone # of __		Request Deputy Chief	Provide Cover for Units
10-89	Bomb Threat		Request Additional Chief	
10-90	Bank Alarm at __		Transfer Fire Alarm Wire	
10-91	Pick Up Prisoner / Subject		Check Fire Alarm Box or Master Box	Hazard
10-92	Improperly Parked Vehicle		Fire Alarm Circuit Open or Trouble on Circuit	

Code	General Purpose	APCO	Norfolk, VA	Walnut Creek, CA
10-93	Blockade		Fire Alarm	
10-94	Drag Racing		Request Gas or Diesel Fuel	
10-95	Prisoner / Subject in Custody		Grass or Trash Fire	
10-96	Mental Subject		In Quarters	
10-97	Check (Test) Signal		Signal Weak	Arrived at Scene
10-98	Prison / Jail Break		Signal Good	Completed Assignment
10-99	Wanted / Stolen Indicated		Fireman Need Help	
10-101	What is Status? (Are you secure?)			
10-106	Secure (Status is secure)			

Index

accuracy, 11-15, 18, 19, 25, 26, **35**
antenna alignment, 5-7, 25, 35
audio tone, 17, 18, 20, 21, 27, 39, 47

batching, 43
beamwidth, 23
beam alignment, 5-7, 25, 35
beam spread, 23, 45
blind zone, 24, 25, 28, 29, 30, 36

calibration, 11-14
Certificate of Calibration, 11-12
configurations, 1
conversions, 16
cosine effect, 7, 23-25, 28, 29, 32, 34, 41

displays, 4
Doppler, 15, 16, 23, 44, 46

factory records, 11-12
false speed, 10, 22, 23, **27**, **37**, 39, 42, 43
fan filter, 9, 18
field disturbance sensors, 38
fixed mounted radar, 2
frequency bands, 16, 44-45
frequency spectrum, 45, 47-48

hand-held radar, 1

interference, 4, 10, 22, 28, 38-39, 44, 45
Interstates, 50

K band, 44, 45
Ka Band, 16, 44, 45

laser radar, 28, 49

maintenance records, 11
microwave, 27, 28, 38, 44 46, 49

modes, 17-19, 28-33

moving mode,

multi unit, 2
mutual interference, 38
moving mode, 19, 30-32, 48

noise, 22, 23, 39, 43

operation, 26-33
opposite direction mode, 20, 30-32

paperwork, 11-14
patrol car transmitters, 39
patrol speed shadowing, 7, 8, 30, 41-42
power source, 1, 10

radar gun, 1
range, 10, 39, 44-45, 48
RFI, 39

same-lane mode, 21, 32, 48
sample time, 25, 35
scanning error, 43
setup, 28
self-test, 17, 26
speaker, 4
spectrum, 45, 47, 48
speed bumping, 43
stationary mode, 3, 17-18, 22, 27-29, 35, 38, 47

tone, 3-4, 14-16, 19
test, 15-21, 26
track history, 27
transmit turn-on time, 40
tuning fork, 3, 11-16, 18-21, 26
tuning fork speed, 14

weather, 22, 44, 45

X band, 44-45

NOTES

NOTES